MEASURING WEIGHT

by Meg Gaertner

Cody Koala

An Imprint of Pop!
popbooksonline.com

abdobooks.com
Published by Pop!, a division of ABDO, PO Box 398166, Minneapolis, Minnesota 55439.

Printed in the United States of America, North Mankato, Minnesota

102019
012020

THIS BOOK CONTAINS RECYCLED MATERIALS

Cover Photos: Shutterstock Images, dial, scale
Interior Photos: Shutterstock Images, 1 (dial), 1 (scale), 16 (spring), 17 (spring), 21; iStockphoto, 5 (top), 5 (bottom left), 5 (bottom right), 6, 7, 9 (top), 9 (bottom left), 9 (bottom right), 10, 13, 14, 19; Red Line Editorial, 16 (box), 17 (box and feet)

Editor: Meg Gaertner
Series Designer: Jake Slavik

Library of Congress Control Number: 2019942418

Publisher's Cataloging-in-Publication Data
Names: Gaertner, Meg, author.
Title: Measuring weight / by Meg Gaertner
Description: Minneapolis, Minnesota : Pop!, 2020 | Series: Let's measure | Includes online resources and index
Identifiers: ISBN 9781532165580 (lib. bdg.) | ISBN 9781532166907 (ebook)
Subjects: LCSH: Weights and measures--Juvenile literature. | Size and shape--Juvenile literature. | Measurement--Juvenile literature. | Mathematics--Juvenile literature. | Scales (Weighing instruments)--Juvenile literature.
Classification: DDC 530.813--dc23

Hello! My name is

Cody Koala

Pop open this book and you'll find QR codes like this one, loaded with information, so you can learn even more!

Scan this code* and others like it while you read, or visit the website below to make this book pop.
popbooksonline.com/measuring-weight

*Scanning QR codes requires a web-enabled smart device with a QR code reader app and a camera.

Table of Contents

Chapter 1

What Is Weight?

All objects are made of matter. Matter is anything that takes up space and has mass. Mass is the amount of matter in an object. Mass and weight are related.

Watch a video here!

Weight is a measure of how heavy something is. Elephants are heavy.

They weigh more than light objects such as leaves.

Mass and weight are similar. But they are not the same. Weight depends on **gravity**.

Chapter 2

Units of Measurement

People in the United States measure weight in units called **pounds** (lb). Small objects might be measured in ounces (oz). One pound is 16 ounces.

Learn more here!

5kg × 20g
11lbs × 1oz

People also measure in **kilograms** (kg). One kilogram is 1,000 grams. It is also approximately 2.2 pounds.

Kilograms are actually a unit of mass, not weight. But people often use kilograms to talk about how heavy something is.

Chapter 3

Tools for Measuring

People measure weight with **scales**. People can stand on a scale to find their own weight. Or they can place an object on the scale.

Learn more here!

d = 100 g, max. 150 kg

There are different kinds of scales. Digital scales have a small screen. Numbers on the screen tell the weight of the object.

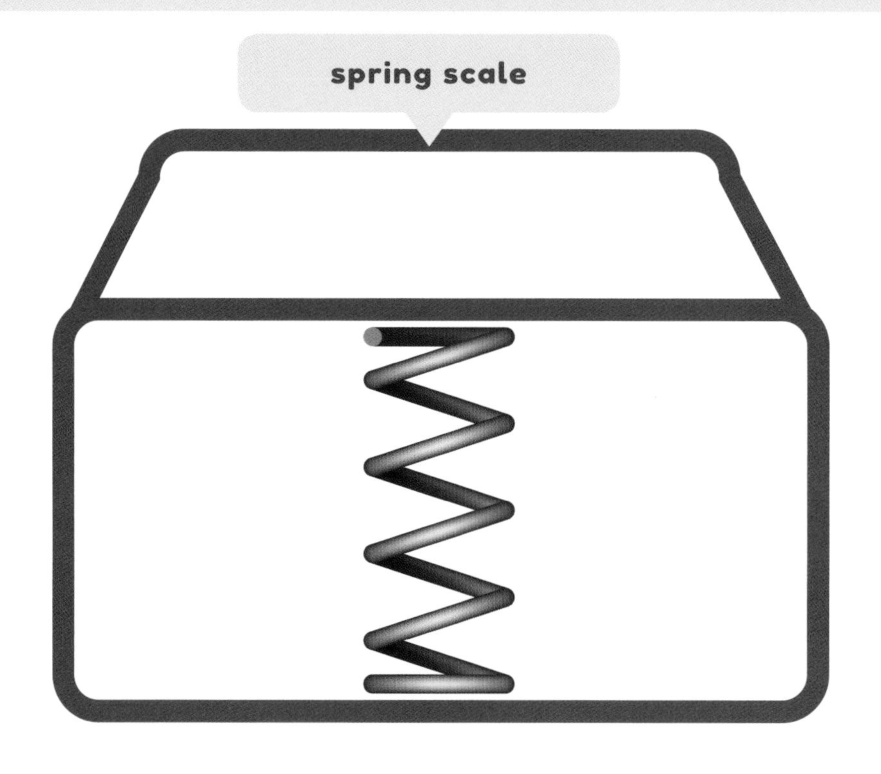

Other scales have **springs** inside. A person stands on a spring scale. **Gravity** pulls the person toward Earth.

As a result, the person's weight pushes down on the spring.

The scale measures how much the spring is pushed down. The front of the scale shows a dial. As the spring is pushed, the dial turns. The dial points to a line or number. It shows the person's weight.

Lines on a spring scale mark the units of measurement.

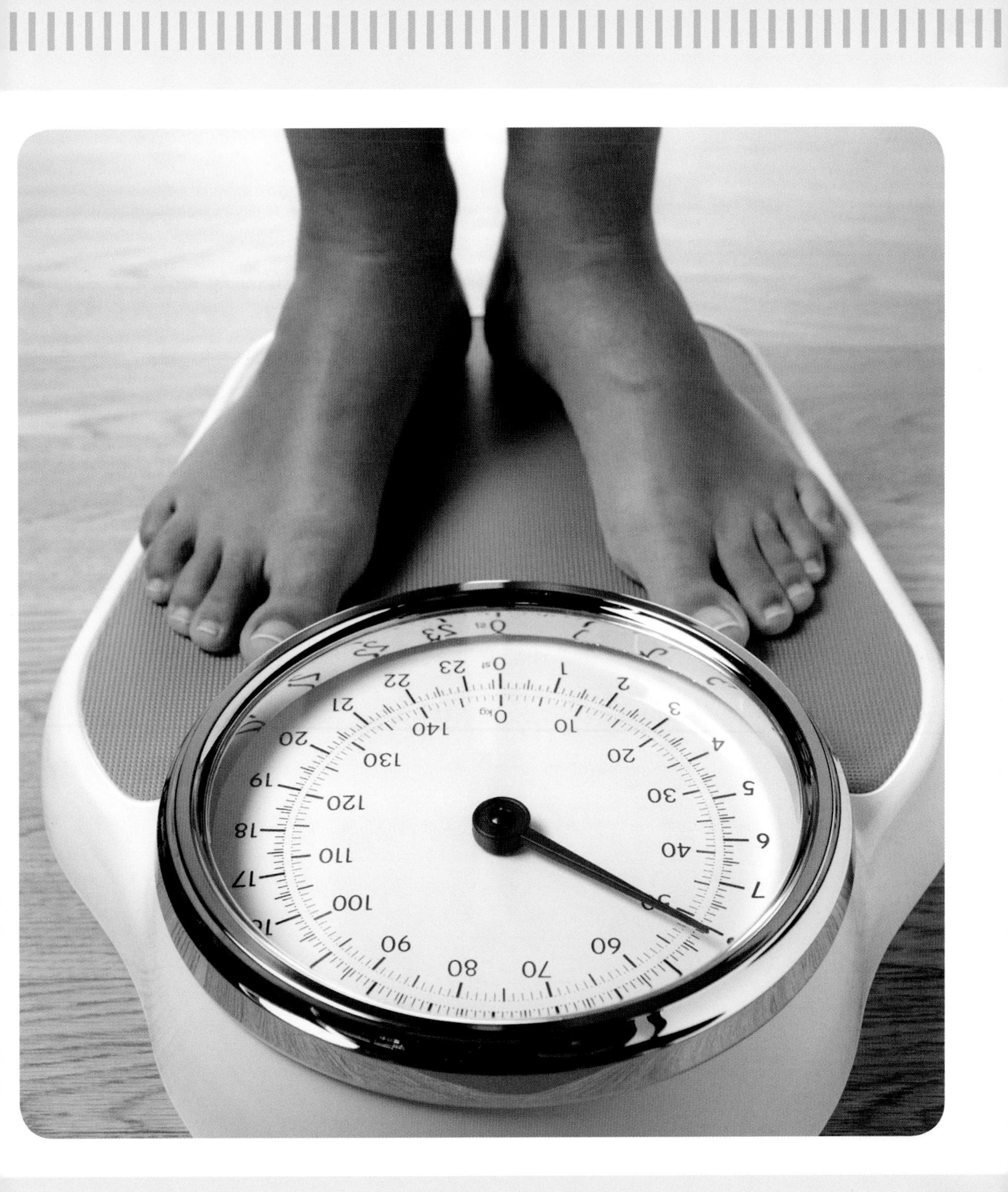

Chapter 4

Measure It!

Nina puts fruit on a **scale**. The dial swings. It points at a line. How much does the fruit weigh?

Complete an activity here!

Making Connections

Text-to-Self

Have you ever had your weight measured? What kind of scale did you use?

Text-to-Text

Weight and gravity are related. Have you read other books about gravity? What did you learn?

Text-to-World

Why might it be useful to know the weight of something?

Glossary

gravity – a force that pulls together any objects with mass.

kilogram – a unit of mass that is equal to 1,000 grams.

pound – a unit of weight that is equal to 16 ounces.

scale – a tool used to weigh objects or people.

spring – a metal coil that can be squeezed or pulled but returns to its original shape when released.

Index

Online Resources

popbooksonline.com

Thanks for reading this Cody Koala book!

Scan this code* and others like it in this book, or visit the website below to make this book pop!

popbooksonline.com/measuring-weight

*Scanning QR codes requires a web-enabled smart device with a QR code reader app and a camera.